THE IMPLICATIONS OF PREEMPTIVE AND PREVENTIVE WAR DOCTRINES: A RECONSIDERATION

Colin S. Gray

July 2007

This manuscript was funded by the U.S. Army War College External Research Associates Program. Information on this program is available on our website, http://www.StrategicStudiesInstitute.army.mil, at the Publishing button.

Comments pertaining to this report are invited and should be forwarded to: Director, Strategic Studies Institute, U.S. Army War College, 122 Forbes Ave, Carlisle, PA 17013-5244.

All Strategic Studies Institute (SSI) monographs are available on the SSI Homepage for electronic dissemination. Hard copies of this report also may be ordered from our Homepage. SSI's Homepage address is:
www.StrategicStudiesInstitute.army.mil.

The Strategic Studies Institute publishes a monthly e-mail newsletter to update the national security community on the research of our analysts, recent and forthcoming publications, and upcoming conferences sponsored by the Institute. Each newsletter also provides a strategic commentary by one of our research analysts. If you are interested in receiving this newsletter, please subscribe on our homepage at *www.StrategicStudiesInstitute.army.mil/newsletter/newsletter.cfm.*

First published by the Strategic Studies Institute:
ISBN 1-58487-298-5

This Juniper Grove edition:
ISBN 978-1-60355-085-7

FOREWORD

If RMA (revolution in military affairs) was the acronym and concept of choice in the U.S. defense community in the 1990s, so preemption has threatened to supercede it in the 2000s. The trouble is that officials and many analysts have confused preemption, which is not controversial, with prevention, which is.

In this monograph, Dr. Colin S. Gray draws a sharp distinction between preemption and prevention, and explains that the political, military, moral, and strategic arguments have really all been about the latter, not the former. Dr. Gray provides definitions, reviews the history of the preventive war option, and considers the merit, or lack thereof, in the principal charges laid against the concept when it is proclaimed to be policy.

Dr. Gray concludes that there is a place for preventive war in U.S. strategy, but that it is an option that should be exercised only very occasionally. However, there are times when only force seems likely to resolve a maturing danger.

DOUGLAS C. LOVELACE, JR.
Director
Strategic Studies Institute

BIOGRAPHICAL SKETCH OF THE AUTHOR

COLIN S. GRAY is Professor of International Politics and Strategic Studies at the University of Reading, England. He worked at the International Institute for Strategic Studies (London), and at Hudson Institute (Croton-on-Hudson, NY), before founding a defense-oriented think tank in the Washington area, the National Institute for Public Policy. Dr. Gray served for 5 years in the Reagan administration on the President's General Advisory Committee on Arms Control and Disarmament. He has served as an adviser to both the U.S. and British governments (he has dual citizenship). His government work has included studies of nuclear strategy, arms control policy, maritime strategy, space strategy, and the use of special forces. Dr. Gray has written 22 books, including *The Sheriff: America's Defense of the New World Order* (University Press of Kentucky, 2004), and *Another Bloody Century: Future Warfare* (Weidenfeld and Nicolson, 2005, distributed by Sterling in the United States). In 2006 he published *Strategy and History: Essays on Theory and Practice* (Routledge). His most recent books are *War, Peace and International Relations: An Introduction to Strategic History* (Routledge, 2007), and *Fighting Talk: Forty Maxims on War, Peace, and Strategy* (Praeger, 2007). Dr. Gray is a graduate of the Universities of Manchester and Oxford.

SUMMARY

Preemption has been, and remains, a leading concept of this decade. But despite its ubiquity in public discourse and its policy relevance, it is a source of great confusion. The term is misused, in some cases deliberately one suspects, but it must be admitted that strategic theorists have offered very little worthwhile reading on the subject. This monograph clarifies the meaning of preemption and distinguishes it from prevention and precaution. It critically reviews the principal charges levelled against preventive warfare and uses that analysis to provide at least the bare bones of strategic theory, more strictly of an alternative to theory relevant to such warfare. The analysis concludes with a set of policy and strategy relevant implications for the United States.

Preemption is not controversial; legally, morally, or strategically. To preempt means to strike first (or attempt to do so) in the face of an attack that is either already underway or is very credibly imminent. The decision for war has been taken by the enemy. The victim or target state can try to disrupt the unfolding assault, or may elect to receive the attack before reacting. In truth, military preemption will not always be feasible.

By way of the sharpest contrast, a preventive war is a war of discretion. It differs from preemptive war both in its timing and in its motivation. The preemptor has no choice other than to strike back rapidly; it will probably be too late even to surrender. The preventor, however, chooses to wage war, at least to launch military action, because of its fears for the future should it fail to act now. In other words, the preventor strikes in order to prevent a predicted enemy from changing the balance of power or otherwise behaving in a manner that the preventor would judge to be intolerable. Naturally, the more distant the anticipated menace, the greater the degree of guesswork as to the severity and timing of the danger. A precautionary war is one waged not out of strong conviction that a dangerous threat is brewing in the target state, but rather because it is suspected that such a threat might one day emerge, and it is better to be safe than sorry.

Put in the vernacular, preventive war, the real subject of this monograph, refers to the option of shooting on suspicion. In an age of weapons of mass destruction (WMD), it could be too late to shoot if one waits for suspicion to be verified by hostile behavior. The official American attitude toward preemption has fluctuated between the admirable declarations of principle by two of its outstanding lawyers: Daniel Webster in 1842 and Elihu Root in 1914. Webster insisted that preemptive action is justified only in the event of a threat that is so imminent that there is no time for other measures. Root took the far more expansive and flexible view that preemptive action must be permitted on a timeline that allows the victim state to take precautions. In effect, Root amended Webster by claiming the legality of a decision to wage preventive war in order to forestall the maturation of the menace. This is not preemption, it is prevention.

Contrary to the impression one might derive from the scale and intensity of the legal debate, it happens to be the case that there really is no legal issue about this subject. International law, in the form of the United Nations Charter, recognizes the inherent right of self-defense by states, and it does not oblige a victim state to wait passively to be struck by an aggressor, although it appears to do so—it is a matter of interpretation. In short, preventive action by way of anticipatory self-defense is legal, or legal enough. Understandably, this permissive interpretation of the license granted by the right of self-defense is open to criticism. In effect, it means that there is no legal constraint on a state's right to resort to force.

The author explains that prevention has not only been a common motive for war, actually it has been quite routine. It is difficult to find historical cases of warfare wherein prevention was not a motivator. Wars typically occur for several, even many, reasons. Prevention is nearly always prominent in the cluster of those reasons. The concept of preventive war has an ominous ring to it that is not entirely deserved. Poor historical understanding is the explanation.

In this monograph, the major criticisms and issues bearing upon preventive war are presented, and their merits and demerits are highlighted. Preventive war is charged with being an act of aggression that is illegal and immoral. We find no value in this accusation, at least as a generality. Preventive war is claimed to be feasible only if intelligence is immaculate. Again, we are unimpressed. It is the view of the author that, with a few exceptions, intelligence needs only to be "good enough." To demand perfect knowledge is to prohibit preventive action. The charge that prevention is seen by some people as a panacea, a "silver bullet," is found to have merit. Next, the claim that preventive warfare must be considered in a framework of probable and possible costs, as well as expected benefits, is approved strongly; it is all too easy to allow one's wishes to suppress suggestions of negative possibilities. Preventive war is charged with prejudging the failure of other policy instruments. This is true, but not conclusively so.

Very often, arguments for more time for diplomacy, sanctions, political subversion, and so forth, are really efforts intended to stall a move to military action, rather than serious claims for prospective success. We need to beware of excuses for endless delay regarding the military option. It is claimed that preventive action sets a most undesirable, even dangerous precedent. This may be true, though it is almost certainly exaggerated, since states scarcely need foreign examples to license muscular self-defensive behavior. To the degree to which it is true, one has to respond by acknowledging the probable fact, while not allowing it to serve as a veto on action we deem essential. Finally, it is fairly popular to argue that a policy that favors prevention represents a futile quest for absolute security. This is a very weak argument. To be willing to act preventively need not be an expression of a foolish search for total security. The argument depends upon a glaring non sequitor.

The monograph offers some thoughts and suggestions with respect to prevention and strategic theory. It observes that there is extant no theory of preventive war. Two reasons for this are postulated. First, the concept is political, and therefore escaped the attention of the rational choice analysts who constructed their narrowly military strategic theory in the 1950s and 1960s. Second, preventive war is not a concept akin to deterrence and containment, despite the suggestion to the contrary made by President George W. Bush. Preventive action is regarded in this monograph as an option that very occasionally is necessary. It does not have the character of a reasonably reliable default strategy, as do deterrence and containment. The following are the principal observations and suggestions for improving theoretical understanding of preventive war:

1. Preventive war is simply war, distinguished only by its timing.

2. Since preventive war is simply war, it is already explained adequately by Clausewitz.

3. Preventive war, in common with all war, is a gamble.

4. The "preventor" begins with the advantage of the initiative, but if success is not achieved swiftly and decisively, that advantage will rapidly diminish as the enemy recovers and counterattacks.

5. The assessment of a preventive war option has to be on the basis of cost-benefit analysis (or guesswork).

6. The anticipation of high costs to prevention action need not be a showstopper. It depends upon the value of the stakes.

By way of conclusions, the monograph identifies some key implications for U.S. policy and strategy. These are drawn from the whole body of the enquiry and, admittedly, contain a few controversial items.

1. Preemption is not controversial.

2. To wage a preventive war requires the fortitude to withstand a great deal of criticism, foreign and domestic. As the principal guardian of international order, the United States needs to be willing to brave that criticism and proceed to do what it believes needs doing.

3. The United States does not need, and should not talk as if it has, an explicit doctrine of prevention (not even when it is thinly disguised as preemption).

4. The United States needs to think politically about its resort to force and be prepared to use military power for political purposes.

5. Preventive war requires very good, but not perfect, intelligence. Of course, it has to be good

enough to serve as a guide to action, but a demand for immaculate intelligence could rarely, if ever, be met. It would function as a needless showstopper.

6. Preventive war is not, and cannot be, a doctrine, no matter what officials claim to the contrary.

7. To choose the preventive war option is to gamble on military success, but to be well-justified in retrospect that success will not necessarily need to be 100 percent.

8. Preventive action, for choice, should take the form of a raid, not an invasion and occupation. The United States should not aspire forcibly to remodel alien societies and cultures. The issue is not one of desirability, but rather feasibility.

9. The principal criteria for preventive action comprise the following:

- Force must be the last resort, not temporally, but with reference to the expected failure of other policy instruments.
- The condition to be prevented by force has to be judged too dangerous to tolerate.
- The benefits of preventive military action must be expected to be far greater than the costs.
- There has to be a high probability of military success.

There should be some multinational support, the more the better. But, the absence of such support should not be allowed politically to function as a veto on actions.

THE IMPLICATIONS OF PREEMPTIVE AND PREVENTIVE WAR DOCTRINES: A RECONSIDERATION

> The United States has long maintained the option of preemptive actions to counter a sufficient threat to our national security. The greater the threat, the greater the risk of inaction—and the more compelling the case for taking anticipatory action to defend ourselves, even if uncertainty remains as to the time and place of the enemy's attack. To forestall or prevent such hostile acts by our adversaries, the United States will, if necessary, act preemptively.
>
> George W. Bush, 2002[1]

> Preemption is the big new rule. It was created by 9/11.
>
> Thomas P. M. Barnett, 2004[2]

> [N]ew world orders, as we have seen, need to be policed.
>
> Michael Howard, 2001[3]

INTRODUCTION: A CONFUSED DEBATE

Rarely has a strategic policy issue generated so much heat, yet shed so little light, as has the contemporary debate over preemption. This ancient strategic idea is not difficult to understand and explain, and it can be considered in the context of more than two millennia of historical experience. Nonetheless, the debate over it since September 11, 2001 (9/11), would seem to have been designed to produce maximum obfuscation.

To cite but a few of the all but willful sources of contemporary confusion: (1) the concept of preemption has been misused; (2) the vital character of the political context, and of the U.S. role in it, has not been debated realistically; (3) largely irrelevant legal and moral issues have been given their traditional outing; and (4) stunningly obvious arguments, typically critical of the idea and practice of preemption and prevention, have been advanced as if they were pearls of eternal strategic wisdom. Six years of confusion and nonsense is more than enough. It is the general purpose of this monograph to remove, or at least greatly reduce, the confusion so as to enable preemption and prevention to be debated intelligently, competently, and in terms that should be useful to policymakers and strategists.

We strategic theorists are guilty of failing to perform our primary professional duty. What is that duty? Carl von Clausewitz could hardly have been plainer: "The primary purpose of any theory is to clarify concepts and ideas that have become, as it were, confused and entangled."[4] Preemption is just such an idea. In order to attack the confusion we shall attempt to rescue the concept, and its much more interesting partner, prevention, from attempted intellectual control by lawyers, moralists, narrowly military defense analysts, scholastic political scientists, journalists, and politicians. The dual intention here is to place preemption and prevention where they belong in the catalog of strategic ideas, and to relate that catalog to its political context. In the latter regard, that means the role of the United States with respect to the maintenance and protection of a tolerably secure international order.

There are no new strategic ideas. In study after study, this author has been moved to quote the following observation by Raymond Aron: "Strategic thought draws its inspiration each century, or rather at each moment of history, from the problems which events themselves pose."[5] Notwithstanding its absurd overstatement, the quotation from Thomas Barnett which heads this monograph is accurate in at least one key respect: preemption as a live policy, and hence strategic subject that certainly owes its recent and current topicality to 9/11. It was ever thus with regard to strategic ideas. As history moves onwards, sideways, indeed frequently in a nonlinear fashion, strategic thought, a la Aron's dictum, responds, faint but pursuing. Preemption and prevention taken together—to be defined carefully in the next section—have been at least contenders for "strategic concept of the decade." If that is felt to be too strong a statement, one cannot deny that along with transformation, irregular warfare, asymmetrical warfare, and the return of counterinsurgency (COIN), they have secured a notable place in public discourse. But, unlike those other strategic ideas, preemption, let alone prevention, does not have any definable military content. The concept is a temporal one and it is preeminently strategic, where military behavior and policy meet.

Officials and commentators, by and large innocently, have spread confusion and invited needless debate. In fact, much of the public debate over preemption and prevention has been not far short of ridiculous. To explain, one can no more debate the general desirability or otherwise of preemption and prevention than one can argue about the general advantages and disadvantages of war. The subject is case specific. War is always undesirable, unless the alternative is anticipated to be even less desirable. What has been remarkable has been the degree to which the literature of the preemption debate has succeeded in wrenching its subject out of context. Preemption and prevention are indeed strategic ideas which can be explained and understood in the abstract. But strategy is, above all else, a pragmatic subject and activity. Preemption has no inherent strategic merit or demerit, save in specific historical contexts. Moreover, to repeat a claim that will appear later in this monograph, preemption cannot sensibly be debated as if it were primarily a legal or a moral issue; it is not. Preemption is a political question that is utterly dependent for its feasibility on military prowess.

Although this monograph provides an overdue reconsideration of preemption and prevention, it is important to recognize that the recent and contemporary debate has more to do with arguments over the American role in the world than it does with the alleged virtues and sins of two strategic ideas. While our argument here will discuss preemption as strategic theory and will endeavor to locate and explain it as such, it must also strive to avoid a characteristic and long-standing weakness in American strategic studies. As befits a public and strategic culture that is more apt to praise Clausewitz than to practice his wisdom, the United States continues to demonstrate an uncertain grasp of the connection between policy and military force. This monograph will demonstrate the essential unity of policy and the threat or use of force. Bereft of political context, a debate over preemption or prevention is literally meaningless.

Far from providing a mere footnote to history, those who have spoken and debated about preemption and prevention have sought to address matters of the gravest significance. There is not much that can compete in importance with decisions for war or peace. Moreover, the preemption issue is not about to disappear. This monograph takes a firm stand on preemption, as it does on the all important superior question of the U.S. role in the world, its policy, and its strategy. But, these pages are not a contribution to the debate over the so-called "Bush Doctrine," which appears to have preemption as its centrepiece.[6] Still less is it written to join the fray, belatedly, over the advisability of the invasion of Iraq in 2003. Hindsight is wonderful, ever the strategist's most reliable friend. References to Bush's Doctrine, if such it be, to Iraq, and even to contemporary troubles with the perennial villains in Iran and North Korea, will be strictly for the purpose of illustration and argument. Doctrines come and go. After the next presidential election, the strategic ideas of preemption and prevention may well take a well-deserved vacation from public overexposure, but the political and strategic contexts that yield them their importance will certainly remain. The policy and strategic issues of preemption and prevention are here to stay, whether Americans like it or not. That is why the subject of this monograph matters deeply. Clear yet sophisticated thinking on preemption and prevention, or the reverse, can have the profoundest significance for international order and American security.

This monograph proceeds to: (1) explain and differentiate the meanings of preemption and prevention, and to place them in the recent history of strategic thought and planning; (2) consider the legal and cultural dimension of preemption and prevention; (3) provide some historical context for what otherwise could be an unduly contemporary discussion; (4) present fairly the major arguments for and against preemption and prevention, while not forgetting that the exercise can have only limited utility as an abstract undertaking; (5) develop a theory, for practice (as all strategic theory must be), of preemption and prevention for the 21st century; and (6) nail its colors to the mast and specify implications for U.S. policy and strategy.

DEFINITIONS

The essential first step to clarity in debate and to understanding the issues has to be the correct use of key terms. It so happens, as nearly every scholarly commentator has complained, that the so-called Bush Doctrine of 2002 either deliberately or accidentally misused the concept of preemption.

The President announced a new doctrine that had preemption in the historically organizing role once occupied by containment and deterrence.[7] What is a doctrine? Analyst M. Elaine Bunn explains: "To call preemption a doctrine implies that it is a central organizing principle for marshalling the instruments of national power in support of national objectives and that in relevant cases, action will be taken in accordance with established governing principles."[8] This is a suitably tough standard for a concept to reach before it can be said to comprise the guiding light for a new doctrine. To date, preemption falls some way short of meeting that standard. Rather more troubling, though, is the confusion created by official U.S. misuse of the concept of preemption. It would be reassuring to know that the misuse has been deliberate and calculated. Thus far, official language and strategic behavior does not offer such reassurance. The conceptual debate sparked by official language and behavior has focused exclusively on the term preemption. This concept has been employed promiscuously to encompass any and all cases of the first use of military force intended to beat the enemy to the punch, even when that enemy is nowhere near ready to throw punches. It is time to restore discipline to the conceptual debate and thereby end needless confusion and assist clear thinking about policy and strategy.

Although the official, and much of the public, debate has been content to argue about the advantages and disadvantages of a multipurpose preemption, in fact there are no fewer than three related concepts jostling for recognition and proper employment. These ideas are preemption, prevention, and precaution. There are two underlying questions fundamental to this whole enquiry: (1) Under what circumstances might one strike first? and (2) On what authority might one strike first?

If confident and reasonable answers can be provided to these two questions, most of the confusion in the on-going debate should dissipate. Before we answer those questions directly, however, it is necessary to define terms so that we know what it is that we are talking about. Each of the three concepts introduced immediately above, preemption, prevention, and precaution, has a distinctive temporal meaning. In terms of extent of temporal distance from an imminent threat, these concepts proceed logically in the following order: precaution—prevention—preemption. The concept of preemption has been employed so expansively that it has wholly overwhelmed prevention and precaution in much of the public discourse. Since the "talking heads" and other commentators who seek to shape public opinion are neither usually well-educated in strategic theory, nor inclined to self-discipline in the exact and proper use of strategic ideas, what we find is a confused and confusing discourse on preemption. Let us move promptly to the essential definitions.

Preemption.

Preemption refers to the first use of military force when an enemy attack already is underway or, at the least, is very credibly imminent. During the Cold War, these concepts were widely understood and, in the case of preemption, were adopted by both superpowers. Both sides planned to launch their strategic nuclear forces on receipt of unambiguous warning (relying on dual phenomenology, or more) that they were under attack. Preemption would have been a desperate effort to disrupt, break up, and forestall a large-scale nuclear attack.[9] A preemptive strike would comprise some combination of a launch on warning (LOW) and a launch under attack (LUA). The alternative would have been to launch after confirmation of attack arrival (LAA), the option otherwise known as retaliation. The verbal formula of the Cold War years held that to preempt meant "to go first in the last resort." Given the enormity of the consequences of nuclear war almost regardless of who went first and also given the very tight time lines for launch decisions and safe escape of land-based forces, it is scarcely to be wondered that preemption was an idea, and indeed a war plan option, that was treated with great respect. Preemptive error today could have many dire consequences. But preemption in the 2000s against a regional menace is not going to end life on the planet. In the 1970s and 1980s, by way of contrast, the strategic context was nowhere near so tolerant of possible error. In those decades, neither superpower could afford to be lax in its strategic conceptualization of preemption, in its contingency planning for it, or in its decisionmaking to choose the preemptive option.

The strategic theory, the policy, the strategy, and the plans for the 21st century need to be radically different from those suitable for the Cold War. But, the Cold War era understanding of preemption should retain its authority, even though the stakes, fortunately, are vastly less today. To preempt is to launch an attack against an attack that one has incontrovertible evidence is either actually underway or has been ordered. In such a context, the only policy and strategy question is "Do we try to strike first in order to try to lessen the blow, or do we receive the blow and strike back?" Of course, the most potent military assets of some polities may be so vulnerable to attack that if they do not strike first, they would not even exist to strike second. Preemption is all about self-defense. Indeed, if we define preemption properly, which is to say as the desperate option of last resort prior to receiving an attack that one is absolutely certain is on its way or all but so, it is not really controversial. The classic statement of justification for preemption was issued by U.S. Secretary of State Daniel Webster in 1842. He faulted the British military initiative in attacking the steamboat *Caroline* in 1837 close by Niagara Falls. The boat was conveying men and arms to fuel a rebellion in Upper Canada. In words that could hardly be bettered for their plain meaning and applicability to our missile age, Webster argued as follows:

> Undoubtedly it is just, that, while it is admitted that exceptions growing out of the great law of self-defense do exist, those exceptions should be confined to cases in which the necessity of that self-defence is instant, overwhelming, leaving no choice of means, and no moment of deliberation.[10]

Webster's very demanding standard of imminence of threat was modified critically by Secretary of State Elihu Root in 1914. Root opened the gates to endless debate when he rejected the traditional American view of what would constitute an imminent threat, and instead recast policy so that it would incorporate strategic judgment. Root asserted "the right of every sovereign state to protect itself by preventing a condition of affairs in which it will be too late to protect itself."[11] This is sensible, but it is no longer preemption. In fact, what Root justifies is a policy of anticipatory self-defense. Exercised with self-restraint and on the basis of excellent intelligence, there is a great deal to be said in praise of the Root formula. However, it is no great stretch to interpret that formula as a legal, moral, and strategic license to wage preventive wars of discretion. Root's use of the word "preventing" is all important.

On the Webster definition, preemption is not controversial. Any state finding itself either actually under attack or unquestionably about to be so, has the right, indeed the duty to its citizens, to defend itself as effectively as it is able. In many, though not all, cases, the best mode of self-defense will be a swift first strike in an attempt to limit damage. Given the assumption that an attack is underway, preemption is not really arguable—save perhaps militarily—because the aggressor has already made the decision for war. It is ironic that the defense and foreign policy community has been debating preemption fairly energetically for the past 5 years and more, even though the concept is thoroughly noncontroversial. That is what happens when people are careless with their concepts. What is, and has always been, controversial, is the Root amendment to the Webster formula. Now we must turn to the second key concept, the ever debateable matter of prevention—of anticipatory, preventive self-defense.

Prevention.

Prevention or preventive warfare is the subject that lies at the heart of this monograph, just as it is what the Bush Doctrine meant when it advertised the occasional necessity for preemption. When a state preempts, it has made a choice between the option of receiving the first blow or striking first. The decision for war has been taken out of its hands. Not so with prevention. If one is uncomfortable with the tough and restrictive Webster view of justified preemption, perhaps there can be a strategically prudent alternative that falls short of constituting a hunting license to wage aggressive wars of discretion. In practice, there often is a middle way, while in theory also one can identify guidance for preventive war that should restrict the discretion of fearful policymakers. However, it is the view of this monograph that once the most restrictive meaning of preemption is abandoned, the flood gates to potential policy and strategy abuse are wide open. This is a case wherein the distinction between two closely related strategic concepts is both crystal clear and vitally necessary. We should not tolerate without objection public policy language that confuses preemption with prevention.

Most major strategic concepts have at their core, indeed depend on, an essential insight which, if not fully appreciated, is certain to have unfortunate consequences. For example, it is surprising to realize that many people fail to understand that deterrence only works if the intended deterree chooses to be deterred. Prevention, preventive self-defense, has at its core the proposition that the preventor, if one may coin the term, is able to detect, and to anticipate, deadly menace in the future. How far in the future? Since there is no theory of preventive war, at least there is none known to this theorist, we must carve our own path through the jungle of conceptual confusion. Once the certainty of imminent-or-actual attack is rejected as being far too passive, and therefore imprudent, in an era that is witnessing the proliferation of weapons of mass destruction (WMD), strategic theory is not able to provide much help to policy and strategy. That is not quite true, as we shall demonstrate, because this monograph views policy and strategy as arts, not sciences. Unfortunately, futurology in its several forms—astrology, advanced methods of defense analysis, and the rest—all comes down to guesswork.[12] Of course, guesses are dressed up as calculated risks, foreseeable futures, and so forth, but former Secretary of Defense Donald H. Rumsfeld was very much in the right ballpark when he emphasized the significance of the unknowns and the "unknown unknowns." For the past several years, the Department of Defense (DoD) has privileged the deeply Clausewitzian concept of uncertainty in its view of the future. There has been something of an intellectual backlash against this recent high regard for uncertainty, but for the purposes of this monograph, the concept fully merits its official high standing.[13]

Preventive war, perhaps just a preventive strike, can be viewed as a muscular application of Root's 1914 dictum of prudence. But if, obedient to Root's precautionary logic, a state is determined to prevent "a condition of affairs in which it will be too late to protect itself," how much protection should it secure through the use of force? Rephrased, how can a doctrine of preventive war be operationalized? Is it misleading to regard preventive war as a fit subject for a doctrine? We shall return to this topic in a later section. Most powerful strategic ideas are attended by potential pathologies. In the case of preventive war, a leading malady inseparable from it is a quest for absolute security.[14] After all, a policy of preventive war amounts to an unwillingness to live with certain kinds of risk. By preventive action, a state strikes in order to control the dangers in its external security environment, at least that is the intention. Some familiarity with history reveals that the law of unintended consequences is apt to frustrate such attempts, but since when did fearful or overconfident policymakers permit themselves to be deflected from a strongly favored course by caveats derived from historical experience?

The most essential distinction between preemption and prevention is that the former option, uniquely, is exercised in or for a war that is certain, the timing of which has not been chosen by the preemptor. In every case, by definition, the option of preventive war, or of a preventive strike, must express a guess that war, or at least a major negative power shift, is probable in the future. The preventor has a choice. It can elect to tolerate the predicted adverse power shift. Alternatively, it can function grand strategically, and endeavor by, say, diplomatic, economic, subversive, as well as military competitive means, to lessen the growing peril. Obviously, temporally the more distant the danger, the greater has to be the uncertainty. In the early 1990s, Americans were assailed by a fashionable theory that tomorrow's great enemy would be superpower Japan. Less than a decade later, the status of super threat of the future was shared between violent Islamic fundamentalism and China.[15]

To consider preventive war pragmatically, as we must in strategic theory, we have to think in time. To preempt is to act on the basis of certain, absolutely contemporary knowledge. In the sharpest of contrast, to launch a preventive war is to act bereft of temporal discipline. It is probably sensible to maintain that the closer to today is the predicted maturing of danger, the less the risk of unsound prediction. But, are any temporal or other kinds of breakpoints suggested by strategic theory? Does a preventive war doctrine oblige one to consider taking forestalling military action only against states whose capabilities (and intentions?) are estimated to mature within, say, 10 years, or 15 years, or when? And just how great does an estimated threat need to be for it to warrant entry on the preventor's hit list?

Contrary to appearances, perhaps, this monograph is not attempting to ridicule the notion of preventive war. The purpose is strictly explanatory. If one endorses the concept of prevention, there is no evading the kind of difficulty outlined in the paragraph immediately above. Any and every preventive war is launched because its executors believe that it is preferable to fight today rather than tomorrow. But, many wars that were predicted never occurred.[16] Also, many a state or potential coalition that could pose a deadly peril in the future failed to develop in a menacing way. Contingency, personality, surprise, and general uncertainty render strategic futurology a profoundly unscientific enterprise. And the more distant the menace in time, the greater the risk of misestimation. This is not utterly to condemn preventive war as a strategic concept; that would be foolish. But it is to suggest in the strongest possible terms that, as an accepted policy option, it is fraught with an awesome possibility of error. And, perhaps needless to say, if a state wages preventive war against a distinctly immature threat, there will be no way of ever knowing whether the war was prudent or unnecessary.

Precautionary.

Precautionary war, our third strategic concept, is war launched to arrest developments beyond the outer temporal or other bounds of detectable current menace. In other words, a precautionary war is a preventive war waged not on the basis of any noteworthy evidence of ill intent or dangerous capabilities, but rather because those unwelcome phenomena might appear in the future. A precautionary war is a war waged "just in case," on the basis of the principle, "better safe than sorry." It is war most usually located at the far end of the timeline from preemption through prevention in response to an ever more distantly perceived danger. Alternatively, a precautionary war can be launched strictly opportunistically, as an attempt to derive maximum benefit from some more major event. For example, had the United States proceeded from Baghdad to Damascus in 2003, the Syrian option would have been precautionary rather than preventive.

If two of the strategic concepts explained in this section, prevention and precaution, are appreciated to be perilously vague in real world application, at least their meaning should now be plain enough. Although it is vital to achieve strategic conceptual clarity, theoretical neatness and transparency is only one, albeit necessary, step towards prudent policy and strategy. The concept of preventive war has to pass what one can call "the Brodie test." Bernard Brodie wrote:

> Strategic thinking, or "theory" if one prefers, is nothing if not pragmatic. Strategy is a "how to do it" study, a guide to accomplishing something and doing it efficiently. As in many other branches of politics, the question that matters in strategy is: Will the idea work? More

> important, will it be likely to work under the special circumstances under which it will next be tested? These circumstances are not likely to be known or knowable much in advance of the moment of testing, though the uncertainty is itself a factor to be reckoned with in one's strategic doctrine.[17]

The problem today is that one can clarify the meaning of the strategic ideas, as here I hope, but that essential task does not advance the building of the needful theory. There is at present no strategic theory of or for preventive war. This lack is as complete as it is perhaps remarkable, given the all too rich historical record of the phenomenon. A later section will seek, carefully, to add a little useful theory to the bare concept which is all that exists at present.

LAW AND CULTURE

Unlike the obliging certainty of the case for preemption—one is or seen to be under attack—preventive action is nearly always controversial. Even if it is not controversial at the time, should it fail militarily and therefore strategically and politically, it is certain to be the subject of bitter debate. To wage preventive war is to shoot on suspicion. Should the preventive action be intended to forestall developments that require at least several years to mature, then the suspicion, though strong, could hardly offer a compelling reason for war now. If the preventive war is designed to forestall entirely a path of development that would or could be deemed threatening, then one is in the policy realm of precautionary offensive strategic behavior, as explained in the previous section. In truth, the meaning of prevention is as crystal clear as its implications are inalienably uncertain. A preventive war is a war that a state chooses to launch in order to prevent some future danger from happening. Because the future is by definition unknown and unknowable, preventive action has to entail striking on the basis of guesswork about more or less distant threats. And threats, of course, are a matter of guesses about capabilities times political intentions. Capabilities can be predicted with some, one must commit only to some, confidence, but political intentions can alter overnight. To choose to wage a preventive war requires a state to conduct complex cost-benefit guesswork. That exercise is expressed politely as calculation. The state comes to the conclusion that war now is preferable to war tomorrow or, at the least, to an adverse shift in the balance of power.

So far, so clear. But, how is the concept and policy of preventive war to be operationalized? Whereas the preemptor has only two choices, to strike first or to ride out the enemy's first strike and then strike back, the potential agent of preventive war has many choices, at least in theory. A state considering preventive war has a choice of timing, "Should we wait?" Also, if the state is functioning with a national security policy and strategy worthy of being so called, the military option will be only one of the ways in which anticipated evils might be prevented.

Recent and indeed current history demonstrates that when there is a whiff of preventive gunpowder in the air, there will be no shortage of people and states arguing for patience and delay while diplomacy does its job—economic pressures, bribery, and the possible benign effect of domestic change in the preventee, to coin another term. Plainly, the less pressing the strategic case for prompt and hopefully decisive military behavior, the greater the nominal range of alternatives to war. Politicians eager to avoid war, up to a point an understandable and meritorious determination, will never be short of excuses, even good reasons, to postpone hostilities. Who knows what will happen tomorrow? The policymaker who has read his or her Clausewitz will know that war, even preventive war, is always a gamble, and that plans can be upset by the independent will and behavior of the enemy as well as by friction. Moreover, even when policymakers have excellent reason to believe the "victory is certain" briefings by advisors, they should understand that a preventive war, unlike a preventive strike or raid, may not be over when the enemy's regular forces are defeated.[18] Witness Iraq today. A theory, let alone a policy and strategy, of preventive war has to accommodate the implications that there is a lot more to war than warfare. A state and society militarily bested in a surprise assault cannot be assumed to be willing to cooperate with the victorious power of the preventor.

This author holds Clausewitz's *On War* in the highest regard for its probing of the nature of war and strategy. However, *On War*, though a timeless classic, naturally reflects the political and moral context of its time of drafting. The principal Clausewitzian dictum, which holds that "war is merely the continuation of policy by other means," is both true and yet apt to mislead today.[19] War is no longer regarded as just another tool of statecraft. Because of the appalling strategic history of the 20th century, with its two world wars and the longstanding menace of a third, peace has acquired a moral and political value that it did not have in Clausewitz's day.[20] Of course, Clausewitz's dictum retains its authority. War must be waged for political reasons and in a manner that reflects the scope of those reasons. But, war is not "merely" one option along with such others as, for examples, diplomacy, economic sanctions, and political subversion. Ever since 1919 and the founding of the League of Nations in the Treaty of Versailles, war has been morally singularized in statecraft. Unfortunately, this rejection of politically motivated violence has taken root only unevenly around the world. Almost nowhere, save possibly among the older members of the European Union (EU), has it taken command of policy. Nonetheless, *On War*, might misinform people who are considering preventive war as a policy option.

With some good reason, one can claim that the Charter of the United Nations (UN) provides grounds that can justify any use of force. Superficially, the Charter only licences self-defense, which it notes is an inherent right. But it does not restrict what a state may do in self-defense. The Charter, as we explain later, is interpreted widely as not placing a state under the obligation to receive the first blow, or to strike first only on the basis of totally unambiguous warning that an attack is either underway or is about to be launched. In other words, the Charter's recognition of the right and duty of self-defense assuredly can be, indeed is, interpreted as licensing a forestalling blow on the part of the intended victim state. This is not to deny the language of Article 51, which does appear to qualify the inherent right of self-defense with the conditioning phrase, "if an armed attack occurs." But what is an acceptable time lapse between a forestalling strike and the anticipated aggression? Silence meets that question. No legal authority provides an answer. If a state can point plausibly to a truly imminent threat, it is in, or close enough to, the legally and morally uncontentious zone of preemption. However, the UN Charter, in effect, can be, and is, interpreted as tolerating preventive war. Needless to add, perhaps, such an interpretation is as legally sound and politically expedient as it is an obvious violation of the plain nominal, though not genuine political, intent of the Charter. Recall that members of the UN are obliged to forswear the use of force in their international relations, except, of course, in the dire circumstances of self-defense.[21]

Whether or not legal, quasi-legal, and moral issues should count for much in the U.S. debate over prevention is a matter for debate. For the moment we will withhold judgment, being satisfied simply to register the point that war is regarded, nearly universally, as being a qualitatively different instrument of policy from the rest of the tools in the grand strategy basket. This is a fact that Americans ignore or discount at their political peril. The singularization of war as behavior requiring extraordinary justification is by no means strictly the product of 1914-18. The just war doctrine of the Catholic Church has long sought to hold Christians to a fairly tough standard for the resort to war to be legitimate. As we plunge into the argument over preventive war, it is useful to bear in mind the standard six requirements of the doctrine. Just war doctrine requires: (1) a just cause; (2) legitimate authority; (3) right intention; (4) proportionality; (5) likelihood of success; and (6) resort to war only as a last resort.[22] These potent criteria are as unambiguous in their essential meaning as they are useless as a practical guide. When is the last resort? As much to the point, who has the right to decide? According to the UN Charter, every sovereign state has an inherent right of self-defense, and hence has a duty to judge on its own behalf when is its last resort.

On the resort to war, international law is highly permissive in practice. However, it cannot be doubted that, even with its permissiveness duly granted, it does not license powerful states to wage aggressive wars simply because they anticipate a large net benefit as a consequence. They can always find some character of preventive excuse. A preventive war can only be regarded as such if it is waged for the highly plausible specific purpose of forestalling an extraordinary danger. If that standard is relaxed, one is back in the culture of statecraft of the 18th and 19th centuries. In those years wars could be, and were, waged in order to restore a balance of power. For example, the Crimean War of 1854-56 was waged, by Britain at least, for the purpose of curbing the power and influence of Russia. The France of Napoleon III, the principal instigator of the war, was motivated by nothing more serious than a quest for glory for a fragile regime in Paris.

To repeat, there is nothing worth debating about preemption. If the attack is certain, there are only two reasons for withholding the use of force. First, it may not be feasible to preempt. If the attack is already underway, it may be highly uncertain just what remains in known locations of the enemy's forces to be struck without delay. Also, it is a distinct possibility that one's military instrument is not ready to preempt. It may lack the necessary intelligence, or suitable ordnance to inflict crippling damage. Second, it may be judged politically and morally important to allow the enemy to fire the first shot and thereby brand himself unquestionably as the aggressor.[23] Needless to say, the second judgment is most unlikely to be persuasive in the face of a nuclear attack. However, if the nuclear attack is very large, specifically if it is from Russia, the only state other than the United States capable of launching a nuclear attack with many hundreds or several thousands of warheads, it may be calculated that there would be no strategic advantage in preempting by way of LOW or LUA. U.S. preemption would not be able to disrupt or blunt an assault on such a scale.

Preventive war has a suitably ominous ring to it. It is almost a frightening concept. By and large, only strong states might wage it, and who or what can restrict their freedom of policy choice? The difference in time between the menaces targeted by preemption and precaution may well be several years, or even longer, but a highly risk averse great power might decide that prompt military prevention for assured control is better than belated efforts at forcible cure. The great or super power as good doctor of international order could persuade itself that timely force is effective preventive medicine. Recall that the default justification for the resort to war is that war has the ability to resolve dilemmas that prove resistant to all other measures. A generously calculated timely assault must preclude knowledge of whether or not any of the non-military tools of statecraft would have succeeded eventually. We will now leave the zone of strategic theory and abstract argument and examine briefly the historical record of preventive war.

STRATEGIC HISTORY

At some risk of complicating matters unduly, we must observe that history does not provide us with a neat and convenient class of plainly preventive wars. What one discovers on close examination is that most—I nearly said all—wars include a preventive motive on the part of a belligerent, and sometimes on the part of both major belligerents. I must hasten to add that the presence of a preventive war motive does not warrant our classifying the subsequent hostilities as a preventive war. The point is that a preventive motive is likely to be only one urge to fight among many, at least in most cases. In order to wage a preventive war, it is not always necessary to fire the first shot or be the first to declare war. For example, it is not hard to sustain the argument that for the United States, World War II, both in Europe and in Asia-Pacific, was a preventive project. This may seem counterfactual, not to say bizarre to some readers. After all, did not Japan shoot first on December 7, 1941, and did not Germany declare war on the United States quite gratuitously on December 11? In fact, the United States had been waging preventive economic warfare against Imperial Japan for at least 18 months prior to Pearl Harbor. By a progressively tighter, eventually total blockade on oil and iron and steel, beginning selectively in July 1940, Washington hoped to coerce Japan into changing policy course in China, though the most immediate issue was the Japanese intervention in French Indo-China.[24] This was a thoroughly futile venture, since it required Japan to reverse its foreign policy of 50 years and abandon its dream of great power status and influence. Tokyo believed it had no practical choice other than to fight. The point is, that the United States acted from a powerfully preventive motive, and it applied pressure with the economic and financial rather than the military instrument of grand strategy. Neither the United States nor Japan desired war in 1941, but U.S. measures of economic blockade left Japan with no alternative to war consistent with its sense of national honor. The oil embargo eventually would literally immobilize the Japanese Navy. So Washington confronted Tokyo with the unenviable choice between de facto complete political surrender of its ambitions in China, or war.

With respect to Germany, a more subtle statesman than Adolf Hitler could have sought to remain neutral in the Asia-Pacific War. But for reasons that are not relevant to this monograph, he elected to join his Japanese ally, notwithstanding Tokyo's fairly resolute neutrality in Germany's struggle with the Union of Soviet Socialist Republics (USSR). Although Germany declared war on December 11, 1941, the United States had been exercising a hugely partial variant of neutrality for many months. U.S. warships escorted convoys far out into the Atlantic, and they had orders to sink U-boats on sight. In addition, the lend-lease transaction with Britain of bases for ships and materiel was not exactly proper behavior for a neutral state. In 1940-41, President Roosevelt did not have the domestic, hence the congressional, backing for war with Germany. But he had a powerful preventive motive for such a commitment, if and when it became domestically feasible. The President knew that a Third Reich victorious in Europe, possibly in possession of a substantial fraction of the British Royal Navy, would pose a predictably deadly menace in the long term to the United States. We know that he was correct to be fearful. Hitler did intend to move on from his anticipated victory in the East to the conduct of a global struggle with America. From an economically and strategically secure super continental foundation in a conquered Eurasia, Germany would have prepared for a maritime-air conflict with the United States. Even during World War II, Germany was pouring concrete for a great new naval base at Trondheim in Norway on the Atlantic.

Pearl Harbor was a political threat to Roosevelt's policy; because the American people wanted vengeance against Japan, they had no particular quarrel with Germany. Fortunately, Hitler's rather ill-considered declaration of war solved Roosevelt's political problem for him.[25] In an obvious sense, Germany compelled the United States to wage war, though it must be said that Hitler had shown, for him, extraordinary restraint in 1941 in tolerating unneutral U.S. activities on behalf of Britain and then Russia. But in a less obvious sense, the United States waged a preventive war against Germany. It entered the conflict as early as was politically possible, with the timing dictated by Hitler and despite a very powerful domestic "pull of the Pacific." U.S. policy and strategy never wavered from commitment to the agreed Allied principle of "Germany First" as the prime adversary to defeat. Without Hitler's intervention, it would have been extremely difficult for Roosevelt to commit the United States to the war in Europe.

If we wind the historical record back a generation to 1917, we discover an even plainer example of the United States deciding to wage preventive war. Germany's announcement of its third campaign of unrestricted U-boat warfare provided the occasion, the excuse, for the public moral outrage that permitted President Wilson to ask for a declaration of war. However, Germany was not threatening U.S. security in any meaningful sense in 1916 or 1917, despite some foolish meddling in Mexico and in the United States itself with German-American organizations. Not only was the U.S. decision for war in no sense whatsoever preemptive, it was preventive to the point of being arguably precautionary. In short, the United States chose to wage a preventive war as an Associated Power of the Allies. Wilson recognized that a German-dominated Europe must constitute a serious threat to U.S. national security. In its own most vital interest, America had to prevent a German victory and then exploit the anticipated fact of its dominant weight and influence as having been the deciding factor in the war in order to shape the postwar international order.

In truth, the reasons for war are always many and mixed. What this monograph suggests is that considerations of prevention typically play a sometimes greater, sometimes lesser, role. It is rare to find a conflict wherein there is no spore of a preventive motive to be found. With regard to the great Cold War of 1947-89, it is standard to cite the deadly trio of geopolitics, ideology, and personality as combining to produce the fatal brew which resulted in 42 years of nuclear shadowed global menace.[26] But, austerely viewed, in 1946-47 both Washington and Moscow decided to wage preventive nonmilitary conflict. The United States was determined to prevent the USSR from expanding its sphere of control any further, while the USSR was no less determined to prevent the United States from rolling back its hard won gains.[27]

If we turn from great wars to lesser ones, again focussing on the American historical record, the evidence of preventive motivation is overwhelming. To repeat, motives are always several, if not many, but it is difficult to find examples of American warfare wherein intention to prevent future trouble was not a factor. The Civil War was waged to prevent the destruction of the Union. The Spanish-American War was contrived, among other reasons, in order to prevent European colonial powers picking up the remnants of the erstwhile Spanish Empire. America's internal frontier advanced from the Eastern seaboard to the Pacific because of a long series of aggressive, in good part preventive, wars waged against native Americans, Frenchmen, Spaniards, and Mexicans. More recently, the United States chose to fight in Korea in 1950 to prevent the forcible unification of the peninsula, an outcome believed to have dire implications in Japan and for Stalin and Mao's estimation of U.S. resolve.

The U.S. decision to defend South Vietnam was a preventive policy move. In long restrospect, the domino theory looks better and better, when viewed in the historical context of 1965, that is. The deployment of U.S. combat forces to South Vietnam in 1965 was entirely discretionary; there was no direct, or even indirect, threat to the United States. However, it is not difficult to discern the motive of prevention when one applies the Thucydidean triptych of "fear, honor, and interest," to the U.S. policy dilemma in 1965.[28]

To bring the record up to date, both Afghanistan 2001 and Iraq 2003 unquestionably were preventive wars. The Gulf War of 1991 was preventive, at least in the sense that a powerful motive behind U.S. policy in 1990-91 was to prevent Iraq proceeding from its easy conquest of Kuwait to the oil fields of Saudi Arabia and the small Gulf states. Indeed, so prevalent is the motive of prevention as a spur to war that there is good reason to wonder if the concept retains any useful meaning. Rigorously regarded, prevention probably always has to be present as a motivation. When states or other security communities fight, they have to be motivated by an intention to prevent some undesired condition. I do not wish to demolish the utility of the concept of prevention, or of preventive war. However, I must point out that far from being a rare and awful crime against an historical norm, preventive war is, and has always been, so common, that its occurrence seems remarkable only to those who do not know their history.

It follows inexorably from the analysis above that the so-called Bush Doctrine is historically unremarkable, notwithstanding all the excitement that it occasioned in 2002-03. Of course, the historical, political, ethical, and legal contexts have changed over the century, as they must. It is true to claim that to launch a preventive war in the 21st century does require extraordinary justification. According to one interpretation of what passes for international law today, such a decision needs to be justifiable with direct respect to the needs of self-defense.[29] As noted earlier, the more distant the threat is judged to be from maturity, the more difficult it is to provide compelling arguments in favor of forcible preventive action. One does not have to be a thoroughgoing cynic in order to appreciate that the emergence of a full grown menace most likely would be as convincing a potential justification for preventive action as it would be too mature to be arrested definitively: It would be too late. If the menace in question includes the threat of WMD, policymakers should be expected to consider the precautionary principle that prevention is highly desirable, or even essential, if there is no cure or prospect of tolerable recovery.

Prevention and preventive war suffer from a near demonic reputation that, by and large, they do not merit. We have argued that prevention is an entirely usual motive for war, albeit in company with other reasons to fight. Obviously, the concept, perhaps the principle, of preventive military action, is open to abuse. An aggressive imperial or hegemonic power could wage a series of wars, all for the purpose of preventing the emergence of future challenges to its burgeoning imperium. However, that is less than a killer argument. Virtually every useful and necessary strategic concept can be abused by the unscrupulous. It is a fact that states have always waged preventive wars, or at least have waged wars for reasons that assuredly included prevention as a significant element. Moreover, it is a safe prediction that they will continue to do so, despite evolving norms and laws that require justification in terms of the forestalling of a very plausibly predictable or imminent threat to national security.

DEBATING PREVENTION

Historians of the Cold War tell us that the U.S. Government considered and rejected waging preventive war in the late 1940s and early 1950s in order to forestall the growth of the Soviet atomic arsenal.[30] A similar debate occurred in the early 1960s regarding the Chinese nuclear program. After the Cold War, the United States is known to have prepared seriously, in 1994 in particular, for the option of striking at North Korea's secret and illegal nuclear facilities.[31] Today, the North Korean challenge to the Non-Proliferation Treaty (NPT) regime remains, even though Pyongyang uniquely has exited that regime, while it has been joined in the cross-hairs of would-be preventors by the Islamic Republic of Iran. The latter has become the menace of the decade and beyond, and, as a result, the focus of most recent and current U.S. debate over the merit and otherwise in preventive war.[32] This monograph endeavors to avoid debating the Iranian case specifically, because its purpose is to provide a basis for understanding the issues associated with preventive action in many circumstances.

Lawrence Freedman advises that "Prevention can be seen as preemption in slow motion, more anticipatory or forward thinking, perhaps even looking beyond the target's current intentions to those that might be acquired along with greatly enhanced capabilities."[33] That is interesting, but misleading. It blurs what should be the clear distinction between preemption and prevention. As this monograph has claimed repeatedly, there is and should be no debate over preemption. The inherent right of self-defense recognized by the UN Charter does not oblige a victim or target state to receive the first blow passively. A target state may choose to let an enemy strike first, but that has to be strictly a pragmatic judgment.

The Bush Doctrine, so-called, declared what it miscalled a preemptive intention to prevent the world's most dangerous regimes from acquiring the world's most dangerous weapons. This sounded like truly muscular counterproliferation; a noble cause, indeed. Unfortunately, or fortunately, international and domestic politics, strategy, and military operations, combine to provide complexities that are certain to harass and generally threaten to frustrate the bold counterproliferator.

It is necessary to assert the historical authority of context. Even if the United States does have, or were to have, a doctrine of prevention truly worthy of the title, when and where it would be applied in action must depend upon the specific circumstances of the case at issue. President Bush and others have likened the asserted doctrine of "preemption" to such dominant guiding concepts from yesteryear as deterrence and containment. This is both wrong and dangerously misleading. Both deterrence and containment have the signal virtue that they provide a prudent, relatively low risk, default option for policy. In other words, when in doubt, deter and contain. If one elevates preemption, actually prevention, to the policy conceptual heights as the default option, what is one saying? The answer is that, when in doubt, the United States will shoot on suspicion, taking preventive action on the grounds that it prefers to be safe rather than sorry. To quote the ominous prose of the *National Security Strategy* of

2002, "History will judge harshly those who saw this coming danger but failed to act."[34] To which one could be moved to reply that there will be so many dangers anticipated for the future, that the United States might well find itself engaged in more wars than it can afford or conduct effectively.

This section summarizes and presents the debate over preventive war doctrine by means of presenting and critiquing seven broad charges that have been levelled and repeated all but endlessly in recent years. These are not straw targets for easy demolition. Each has merit, some merit at least. The purpose of this discourse is to arm the reader with the principal questions and concerns that will need detailed attention, and some specific answers, in each historical case. The final two arguments are of a more general character. The five debating points are listed below.

1. **Preventive war is an act of aggression.** As such it is both illegal and immoral. Let there be no confusion over the practical meaning of a decision to wage preventive war. Such a decision translates as an unprovoked attack upon another supposedly sovereign state. Of course, there is provocation, but it is not of the kind that carries weight in court. The preventive warrior is provoked by what he believes the intention of the preventee will be at some time in the future. Some commentators seek to provide justification by stretching the usual meaning of preemption. It is argued that prevention is really preemption assessed probabilistically rather than temporally. In other words, on this line of explanation, a preventive attack is preemptive if one is sufficiently convinced that an attack will, or would be forthcoming.[35] The author admires conceptual ingenuity, but is unconvinced by these arguments. A mild version of the probabilistic judgment in lieu of temporal imminence

is a conservatively prudent adherence to what has come to be known as "the precautionary principle" as an approach to risk management.[36] These concepts—precautionary principle and risk management—can serve as politically, morally, and legally, more acceptable terms for preventive war.

Is preventive war illegal? Why is not such a war simply a war of aggression, since it has not been provoked by harmful behavior on the part of the target state? The answer is that there is no international law that truly restricts the use of force. This is not to deny that there appears to be such law, and it is located primarily in the UN Charter. However, all is not as it seems. To quote the highly relevant, if somewhat depressing, judgment of Professor David Kennedy of the Harvard Law School, "Over the years, what began as an effort to monopolize force has become a constitutional regime of legitimate justifications for warfare. There is no doubt that the Charter system of principles has legitimated a great deal of warfare."[37] He proceeds to explain that:

> The Charter came to be read as a constitutional document articulating the legitimate justifications for warfare. Lengthy articles and books were written parsing the meaning of "aggression" and "intervention." Does economic pressure count? The conventional levers of diplomacy—the routine arrangements of commercial life—suddenly seemed arrayed as a continuum with violence. At the same time, it was hard to think of a use of force that *could not* be legitimated in the Charter's terms. It is a rare statesman who launches a war simply to *be aggressive*. There is almost always something else to be said . . .[38]

We have made repeated reference to what international law, most especially in the form of the

UN Charter, does and does not prohibit. It is time to be more explicit. With assistance from the distinguished legal authority, Professor Leslie C. Green, formerly of the U.S. Naval War College, let us specify the legal context for the resort to force.[39]

- The preamble to the UN Charter expresses the determination of "the peoples of the United Nations to save succeeding generations from the scourge of war . . ."
- Article 1 of the Charter states that the first purpose of the UN is "to maintain international peace and security, and to that end: to take effective collective measures for the prevention and removal of threats to the peace, and in conformity with the principles of justice and international law, adjustment or settlement of international disputes or situations which might lead to a breach of the peace."
- Article 2 (4) of the Charter obliges members to "refrain in their international relations from the threat or use of force against the territorial integrity or political independence of any state."
- Article 51 proceeds to the heart of the matter. It affirms the:

> inherent right of individual or collective self-defence if an armed attack occurs against a Member of the United Nations, until the Security Council has taken the measures necessary to maintain international peace and security. Measures taken in the exercise of this right of self defence shall be immediately reported to the Security Council and shall not in any way affect the authority and responsibility of the Security Council to take at any time such action as it deems

necessary to maintain or restore international peace and security."

The contextualizing clause, "if an armed attack occurs against a member of the United Nations," is a minor legal difficulty for the intending preventor. But it is widely held, sensibly enough, not to restrict the anticipated victim to passivity prior to the assault. That would contradict the prudent exercise of the master principle of "the inherent right of individual or collective self defence."

- Chapter VII of the Charter reserves to the Security Council authority in cases of threats to the peace or acts of aggression "to make recommendations or decide what measures shall be taken . . . to maintain or restore international peace and security." Those measures include economic as well as military action (Articles 41 and 42). Of course, Security Council behavior is always subject to potential show-stopping discipline by the exercise of its veto power by one or more of the five Permanent Members.

Professor Green's excellent summary of the legal context of our subject concludes with the flat claim that "[i]t is clear, therefore, that the Charter does not per se declare war to be illegal or merely criminal, but merely a breach of treaty subject to the sanctions embodied in that treaty."[40] Writing as a political and strategic analyst rather than a lawyer, this author must add to Green's professional judgment recognition of the fact that the UN Charter is more a political than a legal document. Moreover, it is a "living" political document. It is a political document in legal form. The more realistic among the UN's founders were not confused on this crucial point, but the necessary clarity has not always been widely shared. Professional lawyers have a culture that commits them to approach world affairs legally. In addition, many people who simply oppose the use of force all but reflexively, are more than happy to seize on the presumed authority of an apolitical and astrategic reading of Charter language to lend legitimacy to their moral convictions.

Some people take the view that a state can only resort to force, in other words shoot first, if the action is strictly preemptive in self-defense, or if warfare is licensed explicitly by a resolution of the Security Council (UNSC). This interpretation of the Charter is broadly rejected, first, in favor of the view that the inherent right of self-defense does not require a state to wait to be attacked before it can take active measures to protect itself. Second, the UNSC does not represent the moral authority of the global community, claims of convenience to the contrary notwithstanding. In practice, it is driven by the balance of influence among five highly self-regarding Permanent Members whose judgments on the legality of, say, U.S. strategic behavior, will have nothing much to do with considerations of law or morality.

Overall, there is no doubt that preventive war is not prohibited by an international law that is interpreted intelligently. With respect to moral judgment, that will rest upon the persuasiveness, or otherwise, of the claims advanced for anticipatory self-defense, and, of course, upon the interests and popular feelings at stake in a conflict. To summarize: (1) preemption is unquestionably legal, it is self-defense in the face of an unfolding and self-evidently imminent threat; (2) preventive war is legal as a forestalling move for self-defense, but, as behavior, it is indistinguishable from the waging of aggressive war. As Professor Kennedy wryly suggests, aggressors always have some excuse for their misdeeds.[41]

2. **Preemption and prevention are only feasible if intelligence is immaculate.** Robert R. Tomes insists that "[P]reemption, to be an effective component of national security strategy, requires exquisite intelligence. It requires deep insights into adversary capabilities and interests, accurate indicators and warning, prescient decision making capabilities, and superior battlefield intelligence."[42]

This is plausible, but overstated. It is agreeable to have exquisite intelligence, but Tomes is in danger of setting the standard so high that it cannot be met. This monograph suggests, contra Tomes, that for preemption and prevention one has to settle for intelligence that is good enough. Good enough, that is, to enable military force to do the job it is assigned. With respect to preemption, although exquisite, well-nigh perfect, intelligence would be desirable, it is likely to be the case that a lower quality of information will suffice to enable the preemptor to achieve a seriously disrupting effect. In fact, one could argue that given the would-be preemptor's choices—to strike first or to be struck first—it almost does not matter how good is the intelligence. One preempts as best one can with the information available. Since it is far too late to prevent the attack, virtually any harm that can be inflicted on the enemy's confidence, plans, and forces, must be welcome.

Intelligence for preventive war, as a discrete preventive strike, is a somewhat, though only somewhat, different matter. The claim that preventive military action against, for example, North Korea or

Iran, is not practicable because the United States and its allies lack near perfect intelligence on those countries' WMD infrastructures, is as popular as it is fallacious. There may be excellent reasons why preventive strikes against North Korea and Iran would be poor ideas, but the absence of truly "exquisite intelligence" is not one of them. If we believe Clausewitz rather than Sun-tzu, we know that war is a chaotic realm of uncertainty and friction, and that intelligence habitually is flawed. Such is the nature of warfare. The point that needs to be emphasized is not that intelligence does not matter, that would be absurd. Rather is the valid point to the effect that intelligence need not be immaculate in order to be good enough. Certainly with regard to a nuclear program that has yet to produce operational weapons, it does not follow that because one lacks reliable information on every facility, a preventive strike must fail. Fail to achieve what? A preventive strike guided by very good, but assuredly not immaculate, intelligence, could and should retard a nuclear program by many years. Such an enforced delay might well be judged a highly satisfactory military outcome. Of course, there is far more to the issue of prevention than strictly military considerations.

One must note in favor of this second claim critical of preventive action, that less than immaculate intelligence could well prove disastrous if the target state has operational WMD, some, even just one or two, of which escape preventive execution. The merit in active missile defense is self-evident for such a case. Thinking back to Imperial Germany and Grand Admiral Tirpitz's "Risk Fleet," history reveals many cases when the development of a new highly potent military capability creates a period of unusual risk, should the intended foreign target of the military program decide to prevent its completion.[43] Immediately prior to World War I, not only was Germany's immature High Seas Fleet potentially vulnerable to destruction by Britain's Royal Navy, but Russia's "Great Program" of railroad and army expansion presented Germany with the certain future that, by 1917, its enemy to the East would be more formidable.[44]

For a more complex example, as early as 1937 Hitler calculated that unless he could wage and win the wars that he needed in order to rule all of Europe by 1943 at the latest, Germany's enemies would have caught up in the armaments competition. What is more, the huge material resource advantages enjoyed by the British and French Empires and by the USSR—discounting the United States as a possible initial enemy—meant that Germany only had a few years wherein the balance of military assets would be to its advantage.[45] History shows that the anticipation of major shifts in the military dimension of the balance of power can be periods of acute peril. Other states may well reason "now or never." Certainly they will consider the argument that since war in the future is judged highly probable, the sooner it is launched, the better.

It should be needless to add that transnational norms about war have changed over time. In 1914, even in 1939-41, war was accepted as an inevitable, if regrettable, fact of international historical life. Today, war, and preventive war in particular, is not regarded globally as an ordinary instrument of policy. This means that no matter how legal a preventive strike may be held to be, to launch a war, unprovoked, is to surrender the moral high ground. The degree to which this would count as a factor in pre-attack policy deliberation, must depend upon the strength of the anxiety that is pulling the state towards exercise of the preventive option.

In addition to the normative violation that preventive action represents today, the quality of intelligence on the target state has to be ever less reliable the further into the future one is peering. Threat equals capability times political intention, as noted already, when the would-be preventor is substituting perceived probability of danger for temporal pressure, thereby arguably stretching the strategic domain of preemption. But how confident can one be that intelligence on a state's political intentions is accurate even for today, let alone for a period literally years in the future? Also, military programs can founder for a host of reasons: change in key decisionmakers, lack of resources, or a shift in the state's security context, to cite only a few.

It follows from this discussion that intelligence must be ever more questionable, the further into the future it aspires to probe. Furthermore, even intelligence on current conditions and activities, with a view to provision of necessary, or highly desirable, targeting data, is certain to be imperfect. These are high confidence generalizations. However, it does not necessarily follow from these points skeptical of the quality of intelligence, that as a consequence effective preventive action is always militarily impractical. Perfection of knowledge and of cultural understanding are worthy goals, but they are not realistic as strict requirements, as essential enablers, for all cases of possible preventive warfare.

The argument here is not that we should abandon any standard of required confidence in our intelligence. To suggest that intelligence needs only to be good enough does not imply a relaxed approach to the vital subject. Rather, the purpose is to challenge the notion that only "immaculate" or "exquisite" intelligence can be good enough. That may or may not be true.

3. **Prevention is not a "silver bullet," a panacea.** Preventive war, especially a preventive strike, can be regarded by incautious commentators who do not respect Clausewitz as a definitive solution to a problem that appears resistant to all other policy measures. The dynamics of debate drive opposing positions further apart. From being seen as a possible answer to a pressing or even distant dilemma, advocacy of the military option all too easily evolves from the status of possible answer to recommendation as the solution. It is well to remember that for prevention even to be a live option for debate, the issue in question has to be a most challenging one. Advocates of military prevention may be correct in their criticism of nonmilitary options, including long-term deterrence and containment. However, just because deterrence is unreliable, which it certainly is, it does not follow that a preventive strike offers the certainty of a satisfactory alternative.[46]

Intelligence is bound to be imperfect. The surprise preventive attack may not achieve surprise; friction and ill luck may impede efficient execution of the assault as planned; and key elements in the target set might escape destruction or even detection. In other words, the military option cannot offer a guarantee of complete success, and incomplete success might amount to failure. Preventive war, though practicable in some cases, cannot prudently be viewed as a "silver bullet," as a panacea. It is not certain to be swift, decisively victorious, and definitive in positive consequences. Strong advocates of the preventive war option could do worse than remind themselves of the moral axiom that "those who live by the sword shall perish by the sword."

4. **Preventive action, even if militarily successful, can only be assessed properly in terms of its consequences.** The familiar axiom that there is much

more to war than warfare alone applies with almost spectacular accuracy to preventive action. The preventive option, as with all other choices in statecraft and strategy, has to be considered in terms of expected benefits and likely costs. It is a mistake simply to compare estimates of benefits with guesses as to costs, because the two columns are not independent of each other. To be specific, should the preventive strike or war prove militarily unsatisfactory, or to have consequences that commit one to protracted warfare after the swift campaign, then the costs of the preventive option will escalate way beyond the scale of the initial calculation. There are no laws of history, but many bold decisions for action intended to resolve a current or anticipated threat have had consequences that were quite unintended and even thoroughly unanticipated.[47]

Because of the complexity of international relations, it is difficult in the extreme to predict the consequences of behavior. A prime attraction of the preventive option is its promise of swift and decisive action to solve a dilemma that appears likely, even certain, to be resistant to all nonforcible means. But preventive action, even if staged just as a raid, is apt to have costs that were not anticipated. When debating prevention, it is essential that the costs side of the ledger be considered, as well as the benefits. Those costs can include loss of political reputation, whether or not the operation is militarily successful, since preventive warfare is always a choice challengeable on political, strategic, legal, and moral grounds. Even states that share the preventor's alarmist view of a development will worry, lest a decision to wage preventive warfare become a habit. Only a faint line divides prudent prevention from an arrogant overreliance upon force. How quickly does the hegemonic power draw and fire its gun? Is it disinclined to allow much time for the other tools of grand strategy to be effective or to demonstrate that they cannot work? Given the invariably controversial character of a decision to launch preventive warfare, is the preventor, no matter how powerful, prepared to withstand international condemnation of its all but unilateral behavior?

There is far more to preventive warfare than just the military action. As a major act of statecraft, forcible preventive behavior has to be assessed beforehand on a genuinely cost-benefit basis. And the costs have to include political consequences, including first and second-order effects. Above all else in importance, though, will be the military and other consequences of the target state's responses to the attack. It is a besetting sin of policymakers and strategists to neglect to take the independent will and capabilities of the enemy sufficiently into account. This persisting peril is never more likely to appear than in a case where a state has decided that a preventive strike, or war, is the solution to its problems. Careful consideration of enemy options, regular and irregular, military, economic, and political, should precede, not postdate, a decision to exercise the preventive option.

5. **Preventive military action prejudges the failure of other instruments of grand strategy.** Policymakers may be convinced that diplomacy, economic sanctions and bribes, and political subversion will not bring the target state to heel. But by definition, a decision for prevention action is a decision not to allow further time to pass wherein nonmilitary tools would be used in an attempt to persuade, pressure, and coerce the adversary into mending its ways. Since there is what amounts to a global norm licensing the use of force

only as the last resort, a decision for military prevention unarguably must violate that standard. Recall the rather tortured argument cited already, to the effect that a decision for prevention allegedly can rest on the substitution of believed high probability for temporal immediacy. This monograph, though not unfriendly to all cases of prevention, finds the probabilistic defense of preventive war to be unsatisfactory. Only a distinctly simpleminded determinist uneducated in the role that contingency plays in history could believe that a relatively distant danger will mature with a probability approaching certainty. History is too rich and complex, as well as liable to deliver one or two of Rumsfeld's unknowns and "unknown unknowns," for one to be sure, really sure, that only military action now can prevent intolerable danger much later.[48]

The main trouble with this broad objection to prevention is that it tends to be employed by those whose principal motive is not to discipline the target state, but rather to preclude any military action. If the prevention school is inclined to be unduly dismissive of the coercive and persuasive value of nonmilitary options, so the anything-but-war school will always try to insist that more time is needed for nonmilitary options to work. For them, there is no last resort. In effect, this fifth debating point comprises a clash of two bad arguments. This study advocates that each case of potential preventive war has to be examined on its merits, though in the context of the points, factors, and caveats, specified in the next two sections.

6. **To wage preventive war, even to endorse it as policy, sets a highly undesirable precedent that encourages the resort to force in international relations.** This claim is true, up to a point at least. Furthermore, to proclaim the necessity for preemption,

as has the United States of recent years, is to imply that war is not only an acceptable instrument of policy, but that it is a fairly ordinary one. In other words, to endorse a doctrine of preemption-meaning-prevention is to challenge the slow and erratic, but nevertheless genuine, growth of a global norm that regards the resort to war as an extraordinary and even desperate measure. A policy that favors military prevention proclaims that it is acceptable to decide coolly and in good time that war is preferable to the conditions predicted for "peace." One can try to argue that a decision to prevent really reflects necessity, but that is not convincing. The truth is that preventive war is a war of discretion. And the world is full of people, not excluding many among Western publics, who would never choose to go to war so long as there was an alternative, virtually no matter how humiliating that alternative might be.

Does a policy of prevention, let alone actual preventive behavior, set a dangerous precedent? In principle, the answer has to be "yes." In practice, some assert there is and needs to be a double standard.[49] To explain, it is claimed (not unreasonably in the view of this monograph), that because the United States has an extraordinary responsibility for maintaining world order, it is permitted to act, indeed sometimes it has to act, in ways that would not be acceptable if practiced by others. The justification is international security. As the principal guardian or sheriff of world order, albeit admittedly self-appointed, the United States must allow itself the policy and strategy to fulfil its unique responsibilities.[50] The taking of occasional preventive action can be necessary if regional order and peace with security is to be protected. Washington should not be impressed by criticism of its preventive war policy by those states that seek to exercise political power without responsibility in the UN.

No matter what international law affirms to the contrary, all states are not sovereign equals. The UNSC is a great power club, as was the Council of the League of Nations before it, and as, explicitly, was the "Concert System" that functioned usefully from time to time between 1814 and the dismissal of German Chancellor Otto von Bismarck in 1890. The purpose, character, and restricted permanent membership of the UNSC underlines the point that the maintenance of world order can only rest on recognition of the realities of power relations and the contemporary norms that generally govern those relations. Given what world order requires of the United States, or indeed of any guardian state or institution, the claim for a special license to use force is not only reasonable, it is essential. If U.S. behavior should set a precedent, that would be too bad. But it would be a price worth paying, if the alternative had to be a world sheriff armed only with blanks.

7. **A policy that favors preventive warfare expresses a futile quest for absolute security.** It could do so. Most controversial policies contain within them the possibility of misuse. In the hands of a paranoid or boundlessly ambitious political leader, prevention could be a policy for endless warfare. However, the American political system, with its checks and balances, was designed explicitly for the purpose of constraining the executive from excessive folly. Both the Vietnam and the contemporary Iraqi experiences reveal clearly that although the conduct of war is an executive prerogative, in practice that authority is disciplined by public attitudes. Clausewitz made this point superbly with his designation of the passion, the sentiments, of the people as a vital component of his trinitarian theory of war.[51] It is true to claim that power

can be, and indeed is often, abused, both personally and nationally. It is possible that a state could acquire a taste for the apparent swift decisiveness of preventive warfare and overuse the option. One might argue that the easy success achieved against Taliban Afghanistan in 2001, provided fuel for the urge to seek a similarly rapid success against Saddam Hussein's Iraq. In other words, the delights of military success can be habit forming.

On balance, claim seven is not persuasive, though it certainly contains a germ of truth. A country with unmatched wealth and power, unused to physical insecurity at home—notwithstanding 42 years of nuclear danger, and a high level of gun crime—is vulnerable to demands for policies that supposedly can restore security. But we ought not to endorse the argument that the United States should eschew the preventive war option because it could lead to a futile, endless search for absolute security. One might as well argue that the United States should adopt a defense policy and develop capabilities shaped strictly for homeland security approached in a narrowly geographical sense. Since a president might misuse a military instrument that had a global reach, why not deny the White House even the possibility of such misuse? In other words, constrain policy ends by limiting policy's military means.

This argument has circulated for many decades and, it must be admitted, it does have a certain elementary logic. It is the opinion of this enquiry, however, that the claim that a policy which includes the preventive option might lead to a search for total security is not at all convincing. Of course, folly in high places is always possible, which is one of the many reasons why popular democracy is the superior form of government. It would be absurd to permit the fear of a futile and dangerous quest for absolute security to preclude prevention as a policy option. Despite its absurdity, this rhetorical charge against prevention is a stock favorite among prevention's critics. It should be recognized and dismissed for what it is, a debating point with little pragmatic merit. And strategy, though not always policy, must be nothing if not pragmatic.

We turn now to the quintessentially practical realm of strategy. The next section examines tersely the feasibility of developing a strategic theory of preventive war.

STRATEGIC THEORY AND PREVENTIVE WAR

I have taught strategic theory and worked as a strategic theorist for 40 years. In all that time, up to and including the present, I have never come across a strategic theory of preventive war worthy of the ascription. The political and technical feasibility of preemption was studied endlessly during the Cold War, with both superpowers electing to attempt it on the basis of unambiguous warning of attack. Preventive war was debated within government from time to time, but, on the evidence publicly accessible today, it never came close to acceptance as policy. As for the strategic studies literature, the cupboard is virtually bare. There is no strategic theory of prevention. If prevention is regarded as a powerful strategic concept similar in function, domain, and possibly even authority to deterrence and containment, the absence of theory would be hard to explain. The reality is that prevention is not a strategic concept akin to deterrence and containment, or to limited war or arms control. This study asserts that there is both a lesser and a dominant compelling reason why strategic theorists appear to have neglected prevention as a strategic idea.

The lesser reason why the library of working strategic concepts is bereft of notable treatments of preventive war is because theorists, when they thought about the topic at all, dismissed it as being a political, not a strategic, subject.[52] Prevention does not lend itself to the kind of rational choice analysis that has been responsible for much of modern strategic theory; deterrence and strategic stability are the leading examples. It lacks a distinctive logical structure. In short, it is not really a strategic idea at all; it is political. The dominant reason for theorists' apparent neglect of preventive war is that such war lacks a distinguishable character. Instead, it is simply war that policymakers decide to wage by way of anticipatory self-defense, to put a generous interpretation on the decision. In other words, when strategic theorists try to come to intellectual grip with the concept of preventive war, they discover that the adjective refers to matters that defy their expertise, while the noun war already is treated more than competently in the theories of war provided by Carl von Clausewitz, Sun-tzu, and Thucydides. The subject is war. When or if policymakers bank on the potency of the adjective, preventive, they neglect at their, and our, peril the eternal nature of war and warfare.

So given the negative judgments just delivered, what can be said by way of an alternative to theory for preventive war? Whether or not a theory is possible, and this author, to repeat, believes that it is not, what does the defense community most need to understand about this controversial idea? What follows is offered as the bare bones of the functional equivalent of a theory of preventive war.

1. **Preventive war is war, and preventive warfare is warfare**. It is not a distinctive genus of war and warfare. The distinguishing characteristics of preventive action are motive and timing, though the former is so well-represented historically that it is not especially useful as a discriminator. Timing is by far the superior marker.

2. **If preventive war is simply war, it has to follow that it cannot require a unique strategic theory for its understanding and guidance**. It must be governed by the same features that characterize all wars and warfare. To understand preventive war, read and reread Clausewitz carefully.

3. **Preventive war is a gamble because war is always such**. Preventive timing and leading motive do not negate the authority of the Clausewitzian judgment that "[n]o other [than war] human activity is so continuously or universally bound up with chance. And through the element of chance, guesswork and luck come to play a great part in war."[53]

4. **The state or other security entity that launches a preventive war starts with an advantage**. It has selected the timing for combat, and it has the initiative. However, those advantages diminish should the war be other than a single campaign. All attacks lose momentum over time, and many adversaries are able to rally, regroup, and counterattack in various ways. It is not usually safe to assume that the victim-preventee will be a helpless target set. Furthermore, even if a preventive regular character of war achieves rapid victory, it has been fairly common in history for that victory to be marred by "the war after the war." At the core of Clausewitz's trinitarian theory of war was his insistence upon the universal salience of the complex and highly variable relations among passion or hatred, chance, and reason. A state's army may be beaten, but its public might not accept that verdict.

5. **When considering preventive military action, the assessment must include anticipated and possible costs, as well as expected benefits**. Policymakers, in common with the rest of us, are ever vulnerable to the censoring effect of their desires and convictions.

6. **A prudent anticipation of high costs should not necessarily be a showstopper for proposals for preventive war**. Cost-free, casualty-free warfare is a fantasy. This is not to deny that Kosovo in 1999 was a casualty-free enterprise for NATO. However, that remarkable historical episode was the exception that proves the rule. If the case for prevention is believed to be compelling, even the certainty of daunting costs of many kinds cannot be permitted to close down the option from live consideration. Each historical case has to be examined on its own terms at the time. There are no metrics, there is no methodology, to which one can delegate the decision to act or not to act.

These six points, and the body of enquiry behind them, enables this monograph to specify some key implications for U.S. policy and strategy. It may be necessary to emphasize that some of these implications express the personal beliefs of the author. Although the claims and recommendations are historically and analytically grounded, there is no denying that in a few cases they are controversial. So be it.

KEY IMPLICATIONS OF THE CONCEPTS OF PREEMPTION AND PREVENTION FOR POLICY AND STRATEGY

1. **Preemption is not controversial**. It is not always feasible or effective, but its inherent desirability cannot be challenged.

2. **To be willing to act preventively requires a determination to proceed in the teeth of much, even great, political opposition.** The United States has to be willing to strike preventively, very, very, occasionally.

3. **The United States should not have, indeed does not need, an explicit doctrine, so miscalled, of prevention (or preemption, meaning prevention).** Its global role as principal guardian of world order requires it to maintain the capability to behave preventively, and to be willing to use it.

4. **In order to approach the preventive war option prudently, the United States has to accept the necessity for using military force for political ends.**

5. **To wage preventive warfare successfully requires very good intelligence, as does warfare of any character.** It does not require immaculate intelligence. A requirement for the best is the enemy of the good enough.

6. **Military prevention is not, and cannot be, a doctrine, let alone the national security doctrine.** It should be regarded as "an occasional stratagem," certainly not "as the operational concept of choice."[54] To go to war, even just to stage a very limited campaign, is to enter the realm of chance, risk, uncertainty, friction, and potentially exorbitant costs. Deterrence is infinitely preferable, if and when it can work.

7. **To endorse the prevention option is to be willing to gamble on military success.** In some cases, the damage required to be inflicted must be close to 100 percent (e.g., if nuclear-armed missiles are the prime targets).[55] But in other instances military perfection would not be necessary in order for the strike or campaign to achieve worthwhile strategic and political goals.

8. **In most cases, preventive military action should have the character of a raid, not an invasion.** Strategists

should be pragmatic. At issue is not the desirability of conquest, enforced regime change, and societal remodelling, but rather their feasibility and costs. A swamp-draining motive behind the preventive option is simply not sustainable. It must meet with a fatal level of political opposition at home in the United States as well as abroad, it is not affordable, and—last but not least—it is not doable. The United States is not capable of remaking culturally alien societies so that they become shining examples of successful American style globalization. And if the job is impracticable, it cannot be sound policy and strategy to make the attempt.

9. **Since this monograph endorses prevention as a rare, but still vitally necessary, option, it judges these to be the most essential criteria for a decision to act**:

- Force must be the last resort, not temporally, but with respect to the evidence-based conviction that the nonmilitary instruments of policy cannot succeed.
- There must be persuasive arguments to the effect that the conditions to be forcibly prevented would be too dangerous to tolerate.
- The benefits of preventive military action must be expected to be far greater than the costs.
- There must be a high probability of military success. The U.S. preventor would be risking its invaluable reputation, after all.
- There should be some multinational support for the preventive action; indeed the more, the better. However, the absence of blessing by the world community cannot be permitted to function politically as a veto.

Because of this author's strong agreement with the judgment expressed, these words of John Lewis Gaddis will close this enquiry:

> Like most other nations, we got to where we are by means that we cannot today, in their entirety, comfortably endorse. Comfort alone, however, cannot be the criterion by which a nation shapes its strategy and secures its safety. The means of confronting danger do not disqualify themselves from consideration solely on the basis of the uneasiness they produce.[56]

ENDNOTES

1. President George W. Bush, *The National Security Strategy of the United States of America*, Washington, DC: The White House, September 2002, p. 15 (hereafter cited as Bush, *NSS*, 2002).

2. Thomas P. M. Barnett, *The Pentagon's New Map: War and Peace in the 21st Century*, New York: Berkeley Books, 2004, p. 261.

3. Michael Howard, *The Invention of Peace and the Reinvention of War*, London: Profile Books, 2002, p. 124.

4. Carl von Clausewitz, *On War*, Michael Howard and Peter Paret, trans., Princeton, NJ: Princeton University Press, 1976, p. 132.

5. Raymond Aron, "The Evolution of Modern Strategic Thought," Alastair Buchan, ed., *Problems of Modern Strategy*, London: Chatto and Windus, 1970, p. 25.

6. Superior contributions to the debate from opposing points of view are John Lewis Gaddis, *Surprise, Security, and the American Experience*, Cambridge, MA: Harvard University Press, 2004; and Robert Jervis, "Understanding the Bush Doctrine," *Political Science Quarterly*, Vol. 118, No. 3, Fall 2003, pp. 365-388; *idem.*, "Why the Bush Doctrine Cannot be Sustained," *Political Science Quarterly*, Vol. 120, No. 3, Fall 2003, pp. 351-377.

7. In his speech at West Point on June 1, 2002, President Bush notably demoted deterrence and containment. He said,

> [f]or much of the last century, America's defense relied on the Cold War doctrines of deterrence and containment. In some cases, those strategies still apply. But new threats

> also require new thinking. Deterrence—the promise of massive retaliation against nations—means nothing against shadowy terrorist networks with no nations or citizens to defend. Containment is not possible when unbalanced dictators with weapons of mass destruction can deliver those weapons as missiles or secretly provide them to terrorist allies.

The President insisted that "we must take the battle to the enemy." And he announced that "[o]ur security will require transforming the military you will lead—a military that must be ready to strike at a moment's notice in any dark corner of the world. And our security will require all Americans to be forward-looking and resolute, to be ready for preemptive action when necessary to defend our liberty and to defend our lives." "Remarks by the President at 2002 Graduation Exercises of the United States Military Academy, West Point, New York." *www.whitehouse.gov/news/releases/2002/06/print/20020601-3.html*, accessed June 1, 2002.

8. M. Elaine Bunn, "Preemptive Action: When, How, and To What Effect?" *Strategic Forum*, Institute of National Strategic Studies, National Defense University, No. 200, July 2003, p. 7. This is the clearest statement of the requirements for a doctrine that I have come across.

9. For an excellent treatment of the hazards of preemption during the Cold War, see Bruce G. Blair, *The Logic of Accidental Nuclear War*, Washington, DC: The Brookings Institution, 1993.

10. Daniel Webster, quoted in Michael Byers, *War Law: Understanding International Law and Armed Conflict*, New York: Grove Press, 2005, p. 54.

11. Elihu Root, quoted in Lawrence Freedman, *Deterrence*, Cambridge, UK: Polity Press, 2004, p. 90.

12. See Colin S. Gray, *Another Bloody Century: Future Warfare*, London: Weidenfeld and Nicolson, 2005, Ch. 1.

13. Uncertainty is highlighted by Clausewitz as one of the four elements that constitute the "climate" of war. See *On War*, p. 104. The Preface to Secretary of Defense Donald H. Rumsfeld, *Quadrennial Defense Review Report*, Washington, DC: U.S. Department of Defense, February 6, 2006, p. v, states that "the President [when he took office in 2001] understood well that we were entering an era of the unexpected and the unpredictable . . ." For a full frontal assault on the Bush Administration's emphasis

upon uncertainty, see Michael Fitzsimmons, "The Problem of Uncertainty in Strategic Planning," *Survival*, Vol. 48, No. 4, Winter 2006-07, pp. 131-146. See also Stephen Fruhling, "Uncertainty, Forecasting and the Difficulty of Strategy," *Comparative Strategy*, Vol. 25, No. 1, January-March 2006, pp. 19-31; and Talbot C. Imlay and Monica Duffy Toft, eds., *The Fog of Peace and War Planning; Military and Strategic Planning under Uncertainty*, New York: Routledge, 2006.

14. In the judgment of Jeffrey Record, "Preemption or prevention have their obvious attractions as contrasted with deterrence, at least when they work. But they carry the risk of encouraging a hopeless quest for total security." Jeffrey Record, *The Specter of Munich: Reconsidering the Lessons of Appeasing Hitler*, Washington, DC: Potomac Books, 2007, p. 121.

15. Recall George Friedman and Meredith Lebard, *The Coming War with Japan*, New York: St. Martin's Press, 1991. Richard Bernstein and Ross H. Munro, *The Coming Conflict with China*, New York: Alfred A. Knopf, 1997, was a safer bet. The threat industry abhors a vacuum. I admit to being a worker in that industry. See Colin S. Gray, "The Quadrennial Defense Review (QDR), 2006, and the Perils of the 21st Century," *Comparative Strategy*, Vol. 25, No. 2, April-June 2006, pp. 141-148.

16. It is worth noting the phenomenon of the self-negating prophecy. Naturally, it is impossible to know why anticipated wars failed to occur. It is amusing to observe the high confidence that many people place in the proposition that "deterrence works." They are right, at least they are often right. But even just a single failure can spoil a whole decade or longer. Negative evidence is notoriously difficult to document.

17. Bernard Brodie, *War and Politics*, New York: Macmillan, 1973, p. 452.

18. Societies have frequently declined to accept a verdict of the battlefield which decreed their states' military defeat. Regular warfare often has been succeeded by a popular insurgency in what can be called the warfare after the war. Americans, with their tendency to contrast war with peace and war with politics, have to learn to come to terms with a historical reality that is more complex than such a binary approach allows.

19. Clausewitz, p. 87.

20. See Howard.

21. The literature on war, morality, and law is huge. These recent books are especially helpful: Leslie C. Green, *The Contemporary Law of Armed Conflict,* 2nd Ed., Manchester, UK: Manchester University Press, 2000; Michael Walzer, *Just and Unjust Wars: A Moral Argument with Historical Illustrations,* 3rd edn., New York: Basic Books, 2000; Byers; David Kennedy, *Of War and Law,* Princeton, NJ: Princeton University Press, 2006; and Alan M. Dershowitz, *Preemption: A Knife That Cuts Both Ways,* New York: W. W. Norton, 2006.

22. A. J. Coates, *The Ethics of War,* Manchester, UK: Manchester University Press, 1997, is an especially fine treatment of just war criteria.

23. Germany declared war on France on August 4, 1914, but the French government was determined that its army should not be the first to violate Belgian neutrality. It was believed in Paris that a first move into Belgium by the Germans would label them unambiguously as the aggressors. See Robert A. Doughty, *Pyrrhic Victory: French Strategy and Operations in the Great War,* Cambridge, MA: The Belknap Press of Harvard University Press, 2005, p. 55. Even Hitler sought to confuse the issue of who was the aggressor on September 1, 1939, when the SS staged phoney assaults on a few German border installations and provided some dead bodies (from concentration camps) in Polish uniform at the scenes of the crimes.

24. Akira Iriye, *The Origins of the Second World War in Asia and the Pacific,* London: Longman, 1987, is particularly informative. Also see M. Barnhart, *Japan Prepares for Total War: The Search for Economic Security, 1919-1941,* Ithaca, NY: Cornell University Press, 1987; and Colin S. Gray, *War, Peace and International Relations: An Introduction to Strategic History,* New York: Routledge, 2007, Ch. 12.

25. It may be worth noting that, although all historians today agree that Hitler's voluntary declaration of war on the United States was one of his greatest errors, none of Germany's military leaders raised objections. In Berlin, war with the United States was believed to be inevitable and not very important. The war, overall, was expected to be decided in Russia long before America could mobilize a sufficient weight of military muscle to make a significant difference to the course of events.

26. See John Lewis Gaddis, *We Now Know: Rethinking Cold War History*, Oxford, UK: Clarendon Press, 1997; *idem., The Cold War*, London: Allen Lane, 2006; Vladislaw Zubok and Constantine Pleshakov, *Inside the Kremlin's Cold War: From Stalin to Khrushchev*, Cambridge, MA: Harvard University Press, 1996.

27. The most aggressive American move was the Marshall Plan, announced on June 5, 1947, by Secretary of State George C. Marshall. The Plan was intended both to refire the cold boilers of European economies ruined by war, and to counter Soviet influence, especially in East-Central Europe. Wisely, Stalin rejected the American Plan and ordered his burgeoning, but still insecure, new imperium in Eastern Europe to reject it also.

28. Thucydides, *The Landmark Thucydides: A Comprehensive Guide to "The Peloponnesian War,"* Robert B. Strassler, ed., New York: Free Press, 1996, p. 43.

29. Sensible judgments are provided by strategist-lawyer Walter B. Slocombe in "Force, Preemption and Legitimacy," *Survival*, Vol. 45, No. 1, Spring 2003, pp. 117-130. Slocombe comments,

> [T]he interesting and much-disputed legal issues of how the UN Charter, in particular Article 51 (reserving the inherent right of self-defence against armed attack) should be interpreted – and what, to a practicing lawyer is the equally important issue of who has the legitimate authority to interpret it authoritatively – are matters more for scholars than practitioners of international relations.

P. 121.

30. Bernard Brodie, *Strategy in the Missile Age*, Princeton, NJ: Princeton University Press, 1959, Ch. 7, is an essential source. Brodie was a participant-observer. Marc Trachtenberg, *History and Strategy*, Princeton, NJ: Princeton University Press, 1991, is outstandingly useful, as is Steven T. Ross, *American War Plans, 1945-1950*, London: Frank Cass, 1996. See also David S. McDonough, *Nuclear Superiority: The "New Triad" and the Evolution of Nuclear Strategy*, Adelphi Paper 383, New York: Routledge, for the International Institute for Strategic Studies (IISS), 2006, Ch. 1; and the ever helpful Lawrence Freedman, *The Evolution of Nuclear Strategy*, 3rd edn., New York: Palgrave Macmillan, 2003, especially Ch.3.

31. See Robert S. Litwak, "The New Calculus of Preemption," *Survival*, Vol. 44, No. 4, Winter 2002-03, especially pp. 64-65.

32. There are many contextual differences between the North Korean and Iranian cases. Not only does North Korea have potent WMD and conventional military forces with considerable deterrent value, but also it is not an ambitious, rising, regional power. Whereas Pyongyang's WMD are primarily in the service of a defensive and commercial statecraft, those being sought energetically by Tehran would have the political potential to undergird a bid for regional hegemony, as well as to deter U.S. intervention. As always, the U.S. defense analysis industry responds to policy demand. For a good relevant example, see Henry Sokolski and Patrick Clawson, eds., *Getting Ready for a Nuclear-Ready Iran*, Carlisle, PA: Strategic Studies Institute, U.S. Army War College, October 2005.

33. Lawrence Freedman, *Deterrence*, Cambridge, UK: Polity Press, 2004, p. 86.

34. Bush, *NSS*, 2002, p. 84.

35. See David Luban, "Preventive War," Unpublished Paper, prepared for roundtable discussion at the American Political Science Association Conference, August 2003, p. 27.

36. Michael Williams, "Revisiting Established Doctrine in an Age of Risk," *The RUSI Journal*, Vol. 150, No. 5, October 2005, pp. 48-52, points out that "the precautionary principle" is cited in more than 14 multilateral agreements. Williams is not entirely to be trusted on his use of concepts, however, since he writes: "Precautionary war, better known as preemptive war, . . ." p. 51.

37. Kennedy, p. 79.

38. *Ibid.*, pp. 79-80 (emphasis in the original).

39. Green, pp. 9-10.

40. *Ibid.*, p. 10.

41. Kennedy, p. 80.

42. Robert R. Tomes, *US Defense Strategy from Vietnam to Operation Iraqi Freedom: Military Innovation and the New American Way of War, 1973-2003*, New York: Routledge, 2007, p. 141.

43. Tirpitz's ostensible rationale for his High Seas Fleet was that eventually it would be so powerful, albeit still inferior to

the British Royal Navy (RN), that London would not dare risk challenging it in battle. The damage the German Fleet would inflict in its inevitable defeat would leave the RN inferior and vulnerable to other great navies. It is safe to assume that Tirpitz's real objective was to achieve a High Seas Fleet second to none, not a second-class "risk fleet." For his theory, and policy and strategy, to work, the British were required somehow not to notice the Germans coming, since, if they did, they could easily outbuild Tirpitz's planned force. The High Seas Fleet was a multidimensional disaster for Imperial Germany. Its existence and progress guaranteed British hostility, while it siphoned off scarce raw materials and superior manpower of which the army was in great need. See Holger H. Herwig, *"Luxury Fleet": The Imperial German Navy*, London: George Allen and Unwin, 1980.

44. The Russian "Great Program" that was decided on in 1914 planned a 40 percent increase in the size of the army by 1917 and, thanks to French finance, railroad expansion that should accelerate their pace of mobilization and military deployment by 50 percent, again by 1917. David Stevenson, *Cataclysm: The First World War as Political Tragedy*, New York: Basic Books, 2004, p. 18. Also see Hew Strachan, *The First World War, Vol. I: To Arms*, Oxford, UK: Oxford University Press, 2001, pp. 296-316.

45. Hitler was quite explicit on the need for war before 1943. See "Minutes of the Conference in the Reich Chancellery, Berlin, November 5, 1937" (the "Hossbach Memorandum"), in Louis L. Snyder, ed., *Hitler's Third Reich: A Documentary History*, Chicago: Nelson-Hall, 1981, pp. 268-269.

46. The trials, tribulations, but residual value of deterrence, are discussed in Keith B. Payne, *The Fallacies of Cold War Deterrence and a New Direction*, Lexington, KY: University Press of Kentucky, 2001; Colin S. Gray, *Maintaining Effective Deterrence*, Carlisle, PA: Strategic Studies Institute, U.S. Army War College, August 2003; and Freedman, *Deterrence*.

47. When Serbian military intelligence propelled the hapless innocent, Gavrilo Princip, toward his suicidal moment of destiny in Sarajevo on June 28, 1914, it did not intend to ignite a powder trail to a general war. See Strachan, p. 65.

48. For a relevant discussion of surprise, see Colin S. Gray, *Transformation and Strategic Surprise*, Carlisle, PA: Strategic Studies Institute, U.S. Army War College, April 2005.

49. Luban, p. 4.

50. See Colin S. Gray, *The Sheriff: America's Defense of the New World Order*, Lexington: University of Kentucky Press, 2004.

51. Clausewitz, p. 89.

52. The largely American authored strategic theory of the nuclear age was quite remarkable for its narrowly military logic. See Hedley Bull, "Strategic Studies and Its Critics," *World Politics*, Vol. 20, No. 4, July 1968, pp. 593-605; and Trachtenberg, Ch. 1.

53. Clausewitz, p. 85.

54. Record, p. 121.

55. The deployment of reliable, preferably multilayered, antiballistic missile defenses, should increase tolerance of imperfection in a preventive strike against missiles armed with WMD.

56. Gaddis, *Surprise*, p. 33.

Made in the USA
Monee, IL
04 November 2026

67380496R00039